Roshan Cipriani

I0482398

Living Between The Dimensions-More Reality Change Effects

Copyright 2016 Roshan Cipriani
The spiritual and moral rights of the author have been asserted.
All Rights Reserved
No part of this book maybe used or reproduced by any means
graphic electronic or mechanical including photocopying, recording taping
or by any information storage retrieval system without the
written permission of the author.
ISNB-13:978-1533092472
ISBN-10:1533092478

tABLE oF cONTENTS

Dedication

To the people who are trying to save us from disaster at great personal cost.

Especially to the man in his frustration with the courts who said, "When we are all dead we can sue CERN."

Introduction

I'm Just A Bit Con-CERN-ed

There is a theory that states that collective incorrect recollections are in fact signs of parallel worlds with altered timelines and it was titled by writer and "paranormal consultant" Fiona Broome as "The Mandela Effect." This is due to the collective recollections that many people share in which Nelson Mandela died in prison in the 1980's. Although it seems he was released from prison in 1990 and went on to become the president of South Africa instead.

I am one of those people and as I learned about people equally as shocked as I was, it became more bizarre. There is even a controversy over some children's books called the "*BerenstEin Bears*" which many of us read when we were growing up. However now it seems that reality has been altered so much that the name of the bears and the book title has been changed to "*BerenstAin Bears*." There are many theories and ideas about what has occurred and why. I believe that there is at least one other universe parallel to our own. I have been turning this over in my mind since I first experienced this and from my additional experiences I wrote my first book on this subject, *"Living In A Fractured Multiverse: The Reality Shift Effect."*

Since that time the reality shifts of me as well as numerous other people have only increased.

I have written about the experiments being conducted at CERN and the great probability that the collider is in fact one of many such experiments that are creating the opportunities for just such variances in time and space. According to quantum theory, when elementary particles collide at energies so large that the quantum nature of space/time plays a significant role in the collision dynamics, that no one knows how frequently miniature black hole pairs would be produced in the accelerators. They also do not take into account the interaction of quantum theory and gravity, which while possibly creating variances-- multiple ones in fact, could also collide and fracture each other allowing crossover experiences. Since we do not know the initial quantum state of the universe, but from the observation of other systems that replicate similar conditions it is safe to theorize that that it would be unbounded and continuously growing and expanding. This is in line with the most studied among the proposals for the initial state of the universe and is known as the "no-boundary" proposal of James Hartle and Stephen Hawking. Then there is the theory of "imaginary time", much like "imaginary numbers," which is still the subject of current research.

If this all sounds bizarre to you, please remember that the original proposal of special relativity by Einstein, implicitly treated time as a strictly imaginary coordinate in space.

There is also the fact that imaginary time dimensions appear again in quantum field theory.

So the idea isn't unknown in physics. However I believe that someone has altered something that has impacted our universe and apparently other dimensions as well and they have certainly received our attention.

In our daily lives we cannot truly say that we have never done something one day that we have done hundreds of times before and still asked ourselves whether we are doing it correctly because it seems somehow unfamiliar. Why would we need to question ourselves? Perhaps this is a parallel reality form of resetting a new variation or combining of dimensions and we are picking up the subtle differences without realizing what has occurred.

I believe that things from a similar dimension have crashed into ours and due to this our entire universe has started pulling the other dimensions incrementally (time is imaginary) to remove any differences by some form of absorption and integration. It may be a way to avoid a catastrophe which could lead to the devastation of our present universe and the creation of something else in its place. Perhaps the link with this type of phenomenon and many peoples near death experiences can be the unifying element. According to this theory, maybe we actually were not saved from death; actually dying in that dimension by passing through between the dimensions, within seconds (of imaginary time) and entering this new dimension where we feel and term our experiences as "a near death experience."

Having experienced other alternate memories and realizing that I wasn't alone – and that many other people have shared the same alternate memories was just the starting point in my research. It always comes back to the same question of why do so many people share the IDENTICAL alternate memories about specific events and things if there is nothing to this occurrence? Is it possible that parallel dimensions are experienced the same as time travel from the vantage point of linear perception using this variable known as time(which is said to be imaginary) so that there's no difference between an observer and the observed. It would appear that we will never be able to understand it all - unless we each experience it subjectively for ourselves.

If a change between parallel universes changes physical matter and the interconnection links associated with it, in order to be able to even recognize a difference between the realities of our memories they would have to be outside of the confines of our present dimension so that just by noticing these changes we would actually be "living between the dimensions." I believe it is an indication of moving between perceptions of reality, and that something major happened from your own timeline which changed at the very moment you first realized a difference.

The distinction between the past, present and future is only an illusion, however persistent. –Albert Einstein

Religions throughout the progression of history have stated an acceptance of the existence of parallel worlds adjacent to our own, usually only open to tribal healers, wizards or people endowed with special powers. Apparently often they could walk at will between the barriers of different worlds or dimensions. Almost all civilizations expressed a conviction that there was life in the alternate universes and dimensions. The late Carlos Castaneda wrote that the sorcerer Don Juan Matus held that an area best labeled as a "crevasse between realities" in which the physical world indiscernibly gives way to the spiritual world and other dimensions does exist. Many Eastern cultures believe in a spirit world or parallel universe in which their *devas* or immortals, their subordinates and their challengers all reside.

However today when quantum physicists assert that there may be multiple levels of existence which are not only likely but probable, people just refuse to consider it.

Being teleported is something similar to suddenly being removed to some strange foreign land as if on a cloud. This is nothing new or unheard of as in biblical times many people are reported to safely remove themselves from threat, usually by the assistance of angels, and would be moved from one place to another in the blink of an eye. I believe that only a lesser inclined individual would contemplate that these things are "impossible" in our universe.

Teleportation is interrelated to movement by frequencies since they are simply vibrational energies. As vibration energy at certain frequencies can affect particles causing them to separate or to change shapes, we know that movement is one of its effects. It is only a matter of "time" before various particles can exchange locations spontaneously or with assistance.

Recently scientist proposed the theory called the "many interacting worlds" hypothesis (MIW). This theory confirms the idea that parallel worlds don't just exist, but they also intermingle with our very own DAILY. However they believe that parallel worlds only interact with our world on a quantum level and therefore are not easily detectable. If you have had any of the strange and unexplained circumstance previously mentioned you know it is a lot easier to detect than the scientist are asserting. It was earlier in the year when I first wrote all about what the people at CERN were really up to.

The Large Hadron Collider at CERN is an enormous system of tremendously complicated subterranean concealed tunnels on the French/Swiss border. Here is where the European Organization for Nuclear Research (CERN) contracts physicists to literally crash atoms together at high speeds in an effort to discover the tiniest possible form.

Unfortunately they are also actively trying to create black holes as a portal to other dimensions.

According to Albert Einstein's theory of general relativity, black holes are uninhabitable crevasses of space/time that terminate in a "singularity" or a quantity of infinite density. This would be similar to tumbling into a very deep well that has a bottom, however you can never hit the bottom, you only are pushed into a single point; a point of singularity and of infinite density. In other words black holes are less like holes and more like channels that narrow to a set point. The danger however that is the black hole will still suck in everything that lapses its region. Black holes could be one-way portals to other universes according to Steven Hawking. However he likened passing through one to burning an encyclopedia, ".....the information is not lost, if you keep all the ashes. But it is difficult to read." This is a horrible idea for those of us here on the earth that would rather not end up as ashes in a new universe, even if all our ashes are together.

Since no one knows for certain what is actually on the other side of a black hole, if our earth was swallowed by one, we will most likely be pulverized by its massive gravity as we are being impelled with massive force into a parallel universe. One thing the science clearly asserts is that we will not come out anywhere near the way we went in.

A photon does not need mass to be altered by gravity as the cause of its gravity is not mass, but energy and motion, which a photon positively has. So with a photon, the mass is zero and gamma is infinite. Every force is an interface with two particles that contribute in some way with that force.

As photons never stop, the microscopic bit of spark at the speed of light that is visible in a non-vacuum is the photons bumping into atoms; merging and then being spit back out. Photons are not particles exactly; they are more like particle-waves. They exhibit properties of waves and some of particles but not all features of both. This is very interesting because recently, satellites have revealed two mysterious "black hole whirlpools" in the Southern Atlantic Ocean area. These are extremely potent "vortexes" on the ocean which suck water down into unknown depths. The vortexes which have never been witnessed previously are powerful enough to swallow ships, wreckage and even sea creatures. These vortexes are estimated to be moving 1.3 million cubic meters of water per second. While previously unknown two of these black holes have been observed by physicists from Switzerland and the United States. It would appear that portals are opening with the ocean itself.

CERN scientists recently found unexpected spikes in photon particles, and they are heralding something even bigger than the finding of the Higgs Boson particle. Now they are disclosing that they are indeed hunting for new dimensions and it seems that they have finally discovered the secret to it.

According to FBI disclosure documents interdimensional beings have been in our dimension and documented in the West since at least July 8, 1947. This information is easily found under the freedom of information act and is found in the FBI Vault accessible on line. This however is nothing new. Legend has it that there is an interdimensional vortex in Tibet. The legendary *Shambhala*, and its access point may now be proven to be more than a myth. It is mentioned in the Bon scriptures as a land called *Olmolungring*. While Hindu texts such as *Vishnu Purana* mention the village Shambhala as the birthplace of *Kalki*, the final incarnation of *Vishnu*, so this place is well known in various cultures throughout antiquity.

 However it is what has happened recently in reference to that area that has the U.S. military baffled. An American Military helicopter has vanished along its eight passengers somewhere in the shrouded and rugged Himalayan mountains of Nepal. Even though the helicopter was equipped with a GPS device, radio and an emergency beacon it has vanished without a trace. Not surprisingly it disappeared following a magnitude-7.3 aftershock that shook Nepal earlier which killed approximately 96 people.

The world is still asking the same question over a year later. How can a 21st century military helicopter disappear with all of the cutting-edge tracking technology devices that this helicopter was armed with and still be completely lost?

Did it disappear into the region or dimension known for centuries as Shambala? No one knows exactly.

A portal is an energetic linking between two or more dimensions, and when a portal activates a vortex a cloud or haze will usually appear. They occur where dimensions/realities join together. Traditionally it is said that at times small disturbances of energy in an area can create a wave effect allowing other inter-dimensional beings to take advantage of the imbalance and force their way into our world, or open a trail to let something out of our dimension. Perhaps something like a 7.3 aftershock may have done just that with the "lost" helicopter which may have flown into just such a cloud or haze and disappeared.

When it comes to writing about CERN, and all the abnormalities surrounding reality since its activation, what has always stopped me is my not quite knowing where to "start." Just addressing an issue so enormous and so critical to our perception of reality is quite an undertaking. However since the Large Hadron Collider (LHC) after an almost two year halt and several months of technical reworking, the LHC is back on (June ,2016) and now engaging in experiments at the unparalleled energy of 13 TeV, practically twice the collision energy of its first endeavor, the time to address this situation is now. It is projected to produce up to 1 billion collisions per second and no one knows exactly just what will occur at that time.

The Daily Mail UK newspaper ran a story entitled *Are we all going to die next Wednesday;* referencing the starting up of the LHC at CERN-*June 22, 2016*.

In January 2016 tourist visiting the area near CERN claim that an extraordinary whirlwind type vortex appeared over Switzerland while they were on vacation. What they captured on camera show a bright orb entering the vortex swirl in the sky and then it seems to just vanish entirely along with all the clouds and the vortex itself.

Many people including former physicists from CERN have voiced concerns about the many unknowns attached to these types of experiments. One such concern is of course the possibility of creating "stranglets," which are theoretically small bits of matter that are made up of "strange quarks." These are heavier and unstable relatives of the basic quarks that are the stuff of stable matter.

When I think of *stranglets* in physics, I also reflect on scientist of the past who also wrote couplets and verses in literature. To many people these were also "strange quirks."

Recently I came across one of great interest to this subject. It was written by Nostradamus and is his 44th quatrain.

Migrés, migrés de Geneue trestous,
Saturne d'or en fer se changera:
Le contre Raypoz exterminera tous,
Auant l'aduent le ciel signes fera.

Translated as follows:

All should leave Geneva.
Saturn turns from gold to iron,
The contrary positive ray will (RAYPOZ)
exterminate everything,
There will be signs in the sky before this.

What's quoted above is in reality the 44th quatrain of the ninth "century"; sets of one hundred prophecies into which they were assembled, inscribed by Nostradamus.
We can all wonder about this but right now the only thing that might threaten Geneva, Switzerland would be this LHC at full power at CERN.

Nostradamus also predicted that there would be signs in the sky before that. Quite amazingly CERN was colliding gold ions at the RHIC before they built the LHC, and it is interesting to note that iron is the final element made by stellar nucleosynthesis. Stellar nucleosynthesis is the procedure that comprises the fusion of nuclei due to nuclear reactions that occurs in stars. These nuclei have its place in several elements that are heavier than hydrogen. In extremely large explosions materials become iron.

It could also however also be an astrological time date marker. Nostradamus was himself a famous astrologer, well known for his almanac.

The ancient alchemists all used the same symbols for gold as the Sun, and for iron as the planet Mars. So perhaps it can be recited that "at the time Saturn moves from the Sun to Mars" then this would be a set period in time instead of an alchemical transmutation.

Many people consider that Nostradamus spoke figuratively and was just communicating what he saw in his prophecies and visions as best as he could at that time.

Nostradamus lived from 1503-1566, and since that time spellings vary in the copies of old texts, many words have changed meanings and some words have become totally unclear.

While I have read the Nostradamus text in old French and in English and understand some of it; I do not know enough about the prophecies to draw any definite conclusions. However I think it is something worth considering. In all honesty based on my research, if I were in Geneva, I'd pack my bags and say "Goodbye."

Just consider the power and force of 120 tons of helium under 15-20 atmospheres pressure, with considerable amounts of it in an abnormal superfluid state at critically low 1.9 K temperature, and unprotected in the ring to an 8.2 Tesla magnetic field.

Then perhaps accidentally this altered state of helium combusts owing to the enormous TeV energies and a collision force of 10 TeV, yes it would be unbelievable in its devastation.

There is a high probability that micro black holes (MBHs) will be produced in the LHC. The CERN study specifies that MBHs present no hazard since they will evaporate with Hawking evaporation. However, Hawking evaporation has at no time been tried. In numerous surveys, physicists have projected a non -trivial possibility that Hawking evaporation will not work. Common sense dictates that the more powerful the accelerator will be, the more unexpected and unsafe the events that may transpire. These experiments are a danger - they are interfering with miscalculated qualifiers of matter/energy.

I cannot help but think of the famous painting by Spanish artist Francisco Goya of *Saturn Devouring His Son.* Goya portrays Saturn gobbling up one of his sons. His child's head and portion of the left arm have previously been eaten. The right arm has probably been eaten too, though it could be obscured from view in front of the body and pinned in place by Saturn's thumbs. It is a horribly gruesome painting, done quite well artistically.

According to Roman legend (or the original Greek lore), it had been predicted that one of the sons of Saturn would defeat him, just as he had conquered his father, Caelus.

To avert this, Saturn consumed his children minutes after each was born. His wife Ops finally hid his sixth son, Jupiter, on the island of Crete. She then deceived Saturn by offering a stone wrapped up in his place. Jupiter ultimately deposed his father just as the prediction had foretold.

What if the large rings of Saturn are only an analogy used to describe the complex subterranean concealed tunnels rings beneath CERN? Is this the Saturn that will devour its children there in Geneva?

The safety reassurances that CERN has argued, are lacking since various published works of physicists reveal the inadequate and unacceptable reality of them. CERN's cover-up of pertinent data and neglect from its safety calculations is particularly important given the extent of what is at risk.

I believe this is an issue of world safety and the scientist at CERN are drunk on ego and power and have long since consigned the planet to possible destruction without a care. Our lives are where they have always been; in the hands of God. I do however consider the continuing manipulation of our world and the environment and the many ways it has occurred can eventually reveal to us how it happens.

Chapter One

A Case of the "What If's"

This book is the second in the series on CERN, our reality being altered, the phenomenon known as the Mandela Effect, and Reality Shifts. Yet I have encountered no absolute proofs as to whether or not activities such as what transpires at CERN could be the ***sole*** cause of these "anomalies" or if something completely different is happening simultaneous to their activities. However, I have discovered that experiments which were changing how we define this reality were described prior to CERN and their counterparts' activities. My personal belief is that they are working in tandem with other agencies and are responsible for this and so much more. I would not discount the contributions of Fermi lab (high energy particle physic laboratory) and Brookhaven (a laboratory in New York; active since 1947 with a 2.4 mile radius RHIC - relativistic heavy ion collider) along with the numerous other facilities in various worldwide locations.

 Then the questions that must be considered in a search are these. Could experiments similar to CERN and its type in a nearby universe have created effects in this dimension? Could simultaneous experiments in adjacent dimensions/universes have thrown people from one reality into the exact same spot in an alternate reality?

In a different dimension/universe, if the events weren't consistent, could they possibly produce countless alternatives to the current reality's events? Yet if they were consistent, would they produce mistakes, corrections and physical objects attempting to occupy the same space in different time streams/realities/ dimensions with devastating results?

Perhaps this is that interaction of quantum theory and gravity, which could possibly generate variances-- multiple ones at the same location; since those dimensions would collide and fracture each other allowing crossover experiences.

Sounds unlikely to you? Well, consider what occurred in Las Vegas in 2015. At that time just OUTSIDE the Planet Hollywood Resort and Casino a young woman with her toddler in the car, apparently for no reason drove her car onto a crowd of over forty people. Apparently, she decided to run down a crowd of tourists and local inhabitants without a reason. When detained she maintained that she lost control of the automobile and could not explain just what had happened. Witnesses stated that the car apparently slowed down and then inexplicably accelerated wildly. No one could offer any explanations for her strange behavior. Interestingly the Miss Universe pageant was being held at the Planet Hollywood Resort INSIDE at the time of the crash.

This is the same event at which presenter Steve Harvey; a veteran television personality, announced and helped crowned Miss Colombia as Miss Universe. He then moments later had to remove the crown and instead declare that Miss Philippines was the winner to a stunned and incredulous audience.   It is still unclear whether he was mistaken or just read the information given to him incorrectly, or if it was somehow changed after he read it.

Several nights later Las Vegas was once again center stage, as many people reported seeing a bright light spiraling through the night sky. The night sky over Las Vegas is from one of the most illuminated cities on earth. So in order for that light to have gained attention it would have to be extraordinarily brighter than the Vegas sky and of an apparently long duration. The official explanation was that it was just debris from a Russian rocket. To add to the brightness of the nighttime sky this occurred during a full moon. Of course you would wonder why this did happen in Las Vegas. Well, it just all happens to be in the vicinity of Area 6; another government underground experiment and testing location similar to Area 51. This area inhabits 212 km2 (82 mi sq) between Yucca Flat and Frenchman Flat, across Frenchman Mountain. Only one atmospheric nuclear test was admittedly conducted in Area 6, and that was in 1957. Between 1968 and 1995, there were five underground nuclear tests at this site, two of which involved the synchronized detonation of multiple devices in separate emplacement holes.

Supplementing services near the secured complex are a communications building, numerous radiological sciences and technical services buildings, a fire and first-aid station, and many upkeep and storage buildings. The Area 6 Construction Facilities deliver craft and logistical care. The Device Assembly Facility contains about 9,290mi2 (100,000 ft2) of internal floor space within a Critical Assembly Zone comprising over 22 acres. Yes, it is big enough for another CERN, or a facility similar in scope and intent. There is the possibility that something took place here and there was that similar action in another dimension. So in one dimension/universe a Miss Colombia is still Miss Universe, Steve Harvey does not have to be embarrassed and apologetic and a young mother had an uneventful drive through downtown Las Vegas with her toddler in the backseat.

In the book "The Celestine Prophecies" it basically states that whenever a chance "coincidence" transpires, we should take it as a sign and cooperate with it by following the signal to wherever it may go.

For those people that night in Las Vegas, that specific place and time had certain frequencies align, either on their own or with technological assistance that perhaps created a space between the dimensions that is not typically available. This could be where things are switched or exchanged in a moment.

However from the reactions of all the people involved it seems that there was a change of those circumstances here in this reality/dimension/universe with results that ranged from embarrassing to devastating. My understanding of alternate universes/dimensions/realities is that the more recent the reality shift, the more identical and closer the dimensions.

This would make sense because they would have more similarities. While there seems to be no one tangible process to test these occurrences as each one has so many variables, it may eventually come down to finding the one common factor that we believe must exist.

I find that no matter how much I wish to just ignore this and continue on with "life", I am constantly reminded. Almost every person I have met eventually ends up telling me in some manner about their "close call," or "the time I almost died," followed by how everything in their life "changed after that." Perhaps we have a variation on quantum consciousness and it can merge your "dead" life with another still "living" you in another dimension/universe/reality. Most of the people I have met seem to be unaware of the possibility that they are actually "dead;" at least in their old reality. For me I know that this is not the same time stream dimension/reality/ life that I experienced previously. I can remember exactly when I probably died recently as well.

However, since I've had several of these events, (as I discussed in my first book on quantum physics) I have learned more over time.

From countries in the wrong places, and previously deceased people unaware that they are "supposed to be dead," -- to being unable to write a paragraph some days without being constantly autocorrected; everything is different and strange. Therefore in my thinking, either I switched dimensions or universes or realities, or my original location was greatly altered. If this happens continuously it could have profound implications for understanding ourselves and understanding what being real means.

There could be a reality/ dimension where you exist or have existed through the same sequence of life events with some slight inconsequential disparities that only you would be instantly aware of. You could notice the differences and yet because you could not explain it you shrug it off and choose to just ignore it. Could this happen repeatedly since we "shift reality" without being aware? Could it happen in a moment and the difference is so minor it just occurs unobserved by people around us? Could our consciousness be spread out over multiple universes or dimensions of reality at the same time? Could this be what we see as we sleep and term the experience dreaming? When you dream could it be possible to witness your activities in all possible dimensions or universes where you are at one time?

Perhaps it is like jumping from one version of yourself to another to visit and then to leave disturbing nothing.

Maybe Deja-vu is when we experience seeing information from our consciousness obtained from another universe or dimension which enters our normal waking reality just before the event unfolds in our present reality. There are many people who believe that we are able to travel into other dimensions and realities at will. Many of them claim to be able to do it and to return with knowledge and skills that they did not possess prior to doing this. Just by learning about these types of experiences your present reality becomes questionable.

There are many more apparent slips between the dimensions occurring and as they do physical things, they seem to occur simultaneously in our present reality/dimension. A recent example of just such an occurrence is the six volcanoes that erupted simultaneously around the world on May 24, 2016 in places as diverse as Borneo, the South Atlantic and Indonesia. It's a release of built up energy, but the fact that it is a simultaneous release proposes there is a dangerous factor that may be responsible for this interior rise. This force may be what has caused a violent upsurge of thermal dispersal by catastrophism simultaneously worldwide. Which means that the changes inside the earth's interior were being manipulated externally and all at the same time.

The Large Hadron Collider at CERN could be the cause of the volcano eruptions and multiple earthquakes in those areas. Its high magnetic flux affects the earth's natural magnetic activity. Here in the United States, individuals are carefully observing the super volcano under Yellowstone National Park. In recent years the earth in Yellowstone has been expanding, and many spectators are concerned that what we are observing could lead to a full-blown eruption. To me with CERN's activities it is only a matter of time. Are you starting to get the picture? Some scientists believe that a full eruption at Yellowstone would reduce two-thirds of the United States to being completely uninhabitable, because in the event of a full-scale eruption of Yellowstone, practically the entire northwest United States will be entirely demolished. A full-scale volcanic eruption at Yellowstone would cover practically the whole Midwestern United States with volcanic ash. Food production in America would be nonexistent if this were to occur. Yes , the ruling elite hate America, and will possibly use CERN to generate the Yellowstone caldera artificially because it is possible.

The team of Russian Physicists who assisted in the construction of the Large Hadron Collider for the European Organization for Nuclear Research (CERN), located in Switzerland, reported that an experiment was conducted during the colliders "cool down" phase that has caused what they are calling an "antiquark spree" or has literally, "nailed the heart of Earth".

They apparently lost "control" of this experiment on the South American "anchor" which activated an "unexpected' stream of antiquark's" which first ignited an enormous volcanic eruption in the country of Chile. This then shot through the entire earth has further initiating the disastrous 7.8 magnitude earthquake that took place in China.

So, how could an immense, underground electromagnetic loop disturb countries thousands of miles away? Well when the scientists complete their collision experiments, what happens next?" Where does all that energy go? The energy is then concentrated into what is called a "beam dump". This is a vault buried even more underground and is packed with ingredients intended to captivate and distribute this "beam energy". The power capacities were brought up four distinct times in the phase before the earthquake in Nepal. Instead of using that energy for a collision, it was all injected directly into the ground all 4 sets. These people pushed 16 TW of power into the ground. Let me remind you that CERN has its own power plant, because of the amount of energy required for those experiments. They have admitted that they do not know what will be the behavior of the dumped energy. They also do not know how much energy is actually released at the moment of dumping, or even over how much time this energy would need in order to be absorbed in the dump area. So it is safe to say they have no idea what they are actually doing at all.

Dear reader, would you like to know if the lunatics are playing god today or not? I believe that they are responsible for the deaths of many people and if Stephen Hawking is correct, they will destroy much more than that.

In a preface to a new book he contributed to, which is fundamentally an assortment of speeches given by famous scientists and astronomers called 'Starmus', Theoretical Physicist Stephen Hawking stated clearly his concerns regarding the Higgs Boson, and the experiments at CERN. Hawkins suspects that they are unstable and possibly capable of decay.

"The Higgs potential has the worrisome feature that it might become metastable at energies above 100bn gigaelectronvolts (GeV)," *Hawking wrote.*

The imminent danger of that power potential is that it could end time any time.

The LHC is said to dump 1/4 of the world's current annual energy consumption (20 Terawatts) 1/4 of 20Tw is 250,000,000,000 watts. So 1 Ton of TNT has equivalent energy release of 1162 KW or 250,000,000,000 watts / 1162,000 watts (per ton of TNT) which equals 215,146 Tons of TNT. That means that just 1 LHC beam dump has the equivalent energy release as 215,146 Tons of TNT.

If the CERN mechanism placed energy from the circling particles, with the equivalent of that much TNT four times in succession then earthquakes are not only possible, but quite probable.

Chapter Two

The UN-Holy Science of CERN

The abilities of the scientist at CERN and evidences suggesting that it is being used as a literal TIME ALTERATION MACHINE are also rapidly becoming a reality.

Just last month (June, 2016) the scientists employed at the CERN facility initiated a new test called "Awake" that uses "plasma wakefields driven by a proton beam" to stimulate charged particles. On June 24th, images of some extremely peculiar portal like cloud formations were captured on video in the area just above the Large Hadron Collider at CERN. Is there some type of an association concerning this new "Awake" research and these bizarre clouds seen overhead?

I do not know, but there are scientists all over the earth that are worried that the extraordinary experiments being conducted at CERN could actually open a black hole, destroy the known world or open up a portal to another dimension allowing only G-d knows what to enter this one. It is also interesting to note that that the town in France where CERN is in part located is named "Saint-Genus-Poilly." The word Pouilly originates from the Latin word "Appolliacum." It was here that in Roman times a temple was erected for Apollo, and the legend is that this area is an entrance to the netherworld.

It is remarkable that today CERN is constructed on the exact same location. I first began to notice some very odd and honestly, daunting things about this place that asserts repeatedly to be dedicated to scientific investigation and research alone.

Well first of all I noticed immediately the massive statue of Lord Shiva. Outside the entrance of this scientific facility in Switzerland, a Hindu god dancing on the back of a demon seems a bit out of place. Since he is the god who is known as the Destroyer, it is stranger still especially when he is depicted in the form traditionally known as Nataraja; or the dancing Shiva. It is a well-known sculptural symbol in India and widely used as an emblem of Indian religious culture.

As the Lord of Dance *Nataraja*, Shiva executes the *tandava*; the dance in which the universe is created, preserved, and destroyed. Shiva's long, matted hair, generally up in a knot on the top of his head becomes unfastened during the dance and smashes into the planets. This then causes them to be shaken from their orbit or possibly just destroys them altogether.

Well, if you are still unconvinced that something is a bit strange here there is much more. CERN describes itself as "The Gateway to the Universe," and from an astrological point of view it seems a rather large coincidence that CERN was "reactivated" in alignment to the cosmic movement of the planet Saturn.

In fact, I consider that the scientist at CERN were just waiting until Saturn would be nearer and lined up with the earth and sun. I found this of great interest in understanding the importance of Saturn's positioning in the heavens to CERN.

"This optimal positioning occurs when Saturn is almost directly opposite the Sun in the sky. Since the Sun reaches its greatest distance below the horizon at midnight, the point opposite to it is highest in the sky at the same time. At around the same time that Saturn passes opposition, it also makes its closest approach to the Earth – termed its perigee – making it appear at its brightest and largest. This happens because when Saturn lies opposite the Sun in the sky, the solar system is lined up so that Saturn, the Earth and the Sun form a straight line with the Earth in the middle, on the same side of the Sun as Saturn. May 22, 2015, Saturn was at its closest point to earth. The Moon also occasionally passes in front of Saturn - in an event called a lunar occultation - when seen from various parts of the world. Numerous such events took place in 2007 and 2013-14; the next series of lunar occultation's involving Saturn will take place from late 2018." -**The Naked Eye Planets by MJ Powell**

The Large Hadron Collider is an unbelievable enterprise with an enormous 27 kilometer radius lying underground. It is almost 17 miles in diameter making this an extremely powerful device. However this is not the first time a massive scientific project has had openly religious symbols, names or was intentionally placed in a sacred location.

The Vatican itself set up one of the first celestial observatories in the 16th century which is maintained to this day supplemented with the recent construction of its newest observatory called LUCIFER.

The name stands for "Large Binocular Telescope Near-infrared Utility with Camera and Integral Field Unit for Extragalactic Research, or LBTNUCIFUER or LUCIFER and even LUCI for short" it is a chilled device attached to a massive telescope in the United states located in Arizona.

 Strangely it is stated openly to be named for Lucifer whose name itself means "morning star."

LUCIFER is curiously described on the Vatican Observatory website as "NASA AND THE VATICAN'S INFRARED TELESCOPE CALLED LUCIFER -A German built, NASA and Vatican owned and funded Infrared Telescope… for looking at NIBIRU/NEMESIS."

The first camera, LUCIFER I, was fitted to the telescope in 2010. LUCIFER II is also in the planning stages as of this writing. In 1984, the University of Arizona and the Vatican designated Mount Graham as the location for a complex of eighteen telescopes. The fact that this is a sacred place for the Apache was I am sure quite significant. Situated near the northern boundary of the Chiricahua Apache and Western Apache territories, *Dzil Nchaa Si An*, as it is identified in the Western Apache language, is one of the four holiest mountains in America for the Apache, and considered sacred to the all of the area's Native societies. This is why the land was chosen.

After a protracted legal battle with the Native tribes who objected to the rezoning of their reservation which removed their most sacred location from their ownership, the Vatican was simply awarded the land with all the rights and made it restricted access. The security at this mountain is only matched by the security at Area 51 in Nevada. It has military style guards and warnings posted openly against trespassing and being shot.

CERN's detailed goal from the beginning of the Large Hadron Collider project is to discover the origin of the universe, and capture the smallest subatomic quantum level of particles – in order to find the "God Particle," so using areas of sacred or holy ground is something intentionally done by these "scientific researchers".

One of CERN's ventures within the accelerator is really named "the ALICE Collaboration." This is evidently because it will permit the researchers to discover the core of matter in the same manner that Alice explored Wonderland in the children's story book by falling down a large rabbit hole.

Consider this disturbing quote from CERN 's own website: "... *One of the largest experiments in the world devoted to research in the physics of matter at an infinitely small scale.... You are invited to tumble down the rabbit hole into the wonderland of ALICE.*"

 In the story of *Alice in Wonderland*, it was a white rabbit that tempted her and she fell down into the rabbit hole and was trapped. It is interesting to note that in the story the Rabbit protests that he is "late for a very important date." CERN repeatedly refers to itself as the "Gateway to the Universe."

I discovered other ancient names and terms that are being used for CERN experiments that do not bode well for the earth. They are these:

AEGIS -the shield used by the Greek god Zeus

ALPHA -the beginning of everything; letter A

ALEPH - the Phoenician letter which came from the Egyptian ox's head; the Hebrew equivalent to the letter A

ATRAP - a snare; a deception or a ploy envisioned to capture

DELPHI - Greek mythology it was the navel of the world and also a major location of Apollo worship. The location of an eternal flame to Apollo. The city dedicated to Apollo.

COMPASS - used for navigation and positioning; first created as a device for predictions and fortune-telling by the Chinese

TOTEM - a spirit creature, holy object, or representation that aids as an insignia of a group of native peoples

ISOLDE - a character in an Arthurian story: the Irish princess, Iseult of Ireland. One of the adulterous lovers who are eventually murdered.

ATHENA - Greek goddess of wisdom and the daughter of Zeus. Athena is also called Minerva

AWAKE - awake is the reverse of the state of being unconscious, or asleep.

The LHC apparently also had a magnet used for different experiments whose abbreviation was SATAN. It was the Solar Axion Telescopic ANtenna. - *CERN Libraries, 1999.*

One thing I have always been interested in is the apparently unlimited financing of CERN. Who is real behind the mechanism and the experiments? We recognize CERN has received money from numerous governments across the world. However, a lot of people don't comprehend that the Rockefeller family is the force behind some non-profit charity funding sending all the money donated from the United States right into Geneva, Switzerland itself. The facility took 10 years and around $4.75 billion to create and millions more to operate since it was declared active in 2008.

This means that these people control a great deal of what takes place due to their financing of CERN's setups. This power also allows them to send their own physicists from Rockefeller Institute now called the Rockefeller University. These are some of the best scientists in the world in their particular arenas. They have been active in "discovering new particles" and are continuing unchecked with further "private" experiments, since no one can seemingly oppose them. They have already altered and changed many things with this technology and these are what are now being termed the Mandela Effect. This is a worldwide phenomenon in which people remember something contrary to their present reality/dimension. There have been times when I've experienced something similar to this myself. Having experienced what was termed "a close call" or a "near death experience," multiple times in my life, I realized that quite a few things were different somehow right after each event.

At those times there were numerous subtle "differences" in everything. Recently things have taken a huge leap with many not so subtle differences, making it hard to ignore the obvious changes.

I often feel that I have been living what feels like someone else's life. People are in my life with events that I have no remembrance of interacting in and people that I remembered clearly and distinctly are now completely different.

Scientist state that there are parallel dimensions existing where there are multiples of each person. I think we now are constantly being sucked in and out of dimensions and sometimes we just do not get back in to our original dimension. This would explain having clear memories of things that never occurred in the dimension/reality you are standing in at present.

Recently I was stunned to see a United States map. Having been familiar with its shape throughout my childhood I was shocked by its "deformity." Southern Ontario fits into the former silhouette of the USA seamlessly. The southern slope is so deep into the country it now reaches to the latitude of California's Northern border. Having attended college in Michigan and having made numerous trips from Michigan to New York in those years, I am sure I would have remembered having to drive through Toronto or Ohio to get there.

While I cannot speak for anyone else; let alone everyone else-- this was not true in my past reality/dimension.

Now with many changing phrases and contemporary terms printed in a Bible that had been in my home bookcase shelf for years; I know that something is very wrong in this reality, and it is very bad somehow.

Recently, a severe storm that created numerous waterspouts then came onshore at a beach south of the Cuban capital, injuring 38 people. It destroyed homes and damaged buildings. There was no warning and the waves grew to about 16 feet as people ran inside their homes in terror. There has never been a record of this happening before. What if these waterspouts that occurred are directly linked to the activities at CERN? That same Saturday French-speaking Switzerland was hit with violent hail storms which caused damage to vineyards and around 5,300 homes in Switzerland. There also on Lake Neuchatel, waterspouts formed to the astonishment of witnesses. A water spout is actually a type of vortex or portal that opens up in a body of water. In 2011, following the devastating mega-earthquake and subsequent tsunami, just off the coast of Japan, a giant whirlpool/vortex was seen near Qarai City in Ibaraki Prefecture in northeastern Japan. Within weeks, American scientists discovered two massive whirlpools/ vortexes measuring nearly 400 kilometers in width in the Atlantic Ocean, near the coast of Guyana and Suriname.

I believe that this is all a part of the effects worldwide of the smashing together of subatomic particles hours earlier from experiments commenced at CERN's Large Hadron Collider in Geneva. CERN is quite possibly creating micro black holes all over the earth and it is possible that we are looking at the convergence of our reality/dimension with that of multiple others in rapid succession. It is also possible that it could conceivably cause various weather anomalies in our reality dimension.

This is possible if two distinct and quite different worlds have somehow entered a state of quantum entanglement, and gravitational singularity occurs. This is where any form of mass can be compacted to an adequate density so that it will unavoidably collapse into a singularity or event horizon. This would be a situation akin to the beginning of the Big Bang, or the collapse of star where matter would stream into a space-time curvature, all the dimensions affected would decrease to nothing and density and gravity would become unbounded. Is the LHC a star gate device capable of producing wormholes or portals for time traveling beings as well?

Consider this statement from Sergio Bertolucci, the Director for Research and Scientific Computing at CERN- The Register November 2009:

"Out of this door might come something, or we might send something through it." .... "a very tiny lapse of time, 10-26 seconds, but during that infinitesimal amount of time we would be able to peer into this open door, either by getting something out of it or sending something into it."

CERN's convention clearly says that its exploration should be of a "pure scientific and fundamental character." Additionally, the convention declares:

"The Organization shall have no concern with work for military requirements and the results of its experimental and theoretical work shall be published or otherwise made generally available."

Under the excuse of scientific research, the entire world donates to this religious pursuit, and they are protected by the ultra- clandestine Swiss government. Since the diffusion of responsibility across many nations and employees, it has created a climate of confusion within the whole project. No one identifies who is in charge; no audits can ever be conducted; no backing is ever sufficient; no actual results are ever published. They are in violation of their own published convention. As usual, the majority of humanity has no idea of the true threat of CERN and how it is being used against the earth. The truth is that they are playing God and they do not have the earth's best interest at heart.

The facility of CERN's security is the highest possible, because it is top-secret "science" and no one has given any confirmable tours or have even viewed the underground tunnels. These scientist plan experiments covertly and then only announce the information much later. The statistics looks exactly like the models they created earlier with no variances. Few people have ever actually witnessed a collision, per their own account. No one really knows what CERN is doing; as there are no "real time" action logs.

Other colliders around the world are in clear operation above ground. There is no purpose for placing an atom smasher underground and numerous reasons not to.

So why did CERN need to tunnels 575 feet below ground except to hide exactly what they are doing from the public. When the scientists at CERN do finally release "technical findings" why do they have nothing to do with particle acceleration? Why is CERN releasing its detection of music inhabiting a distinct realm when it is irrelevant? As an example CERN's recent analysis that "classical music exists in the fifth dimension and humans cannot go there" is simply one of the more ludicrous "findings" that the CERN super computers have been employed in.

"No one knows what the thing does – no one does. Firing particles at each other at the speed of light can't end well. We (physicists) have no idea what we're doing."- **Dr. Amit Goswami, Quantum Physicist**

If the full scope of the project was actually known the public would start demanding transparency for a "pure science" setup that is taking trillions of dollars from the people of the world. However you must decide for yourself what you will believe.

Chapter Three

Isn't More Always Better?

Recently I learned that the group at CERN wishes to enlarge its facility by constructing an atom smasher ten times larger than the existing ones. Switzerland will of course be the selected receiver of the monies that will also be ten times bigger than the already current CERN operating expenses of over 1 billion dollars per day. The second, ten times larger loop will be excavated under Lake Geneva and the adjacent region, which is one of the best and most exclusive parts of Europe.

Surprisingly, no one is asking just what type of ecological destruction will ensue when the enlarged tunnels are unearthed under lakes, townships, and the beautiful Swiss landscape. Today CERN is the biggest scientific partnership in the world. Consequently, its integrity by association is unequalled. Now there are 25,000 of the world's best scientists employed by it, yet concerns still exist regarding the careless approval for international physicists to be allowed to attempt such possibly dangerous research with likely dire consequences for the earth and its inhabitants.

I am sure that nothing could be more life-threatening than opening black holes that could eat the earth or discovering anti-matter that extinguishes the entire Universe if it is possible.

Just working on something of that nature is irresponsible at the very least.

And by the way why build something of this magnitude in Switzerland? These people have only built knives, clocks and watches for hundreds of years; it is hardly the hub of modern construction or theoretical physics technology. It takes a lot of imagination to believe that these people have constructed the greatest and most complex scientific machine in the world. The only thing the Swiss do well, and are internationally known for is privacy; especially fiscal privacy. Once that project was located there you can be assured that you will never get a full accounting of anything from CERN; either financial or operational. Even though over ninety percent of the CERN tunnels are located under France, cleverly Geneva is in complete control of CERN funds. A thorough audit of CERN will never be possible due to the private Swiss banking laws and protocols. CERN is completely hidden from sight and as protected as possible in a country that is legendary for protection of secrets.

Some people consider that CERN is a collection and operation center for ECHELON and the CIA's data mining center. This would not be unbelievable since CERN claims leadership in data mining, surveillance, and is currently housing the biggest collection of Cisco routers in Europe.

Cisco Systems, Inc. is a United States based multinational technology corporation located in San Jose, California, that designs, manufactures and vends networking equipment. It is considered to be the main networking enterprise in the world. Cisco sold commercially successful routers first and became known for supporting multiple network protocols, so that their routers quickly became vital to Internet service providers. By the late 1990's Cisco scored a vital domination in this critical sector. Basically, every internet message sent initially went through a Cisco router.

However based on my research I have no way to know if CERN is a Network Access Point for observation for the ECHELON spy system, but it certainly could turn out to be that as well. CERN is one of the original Network Access Points –NAP- because it set up the source code for creating the World Wide Web.

The Hadron Super Collider is just one of CERN's projects. For the purposes of this book I have confined my research to it and its *likely* being one of those experiments contributing to the effects in altering merging or completely destroying dimensions/time streams and realities. Reality shifting means **not** sharing the same perception of reality as the present majority. It means numerous individuals experiencing a different past from the majority.

It is a complex and confusing issue, and when you are the one *not* sharing the perceptions on any or multiple points as the present majority it can come as quite a shock. Now with a bigger mechanism in the planning I anticipate that there would be even bigger, more noticeable changes, for even more people.

Something unexplainable has happened. For numerous people the world seems to have changed. Perhaps when a dimension/ reality becomes discordant with your continued existence; you die there, then your consciousness "merges" to the adjacent useable dimension/reality in which you are still alive, and you continue to live on as you but with subtle differences.

A new side effect of this is that the new dimension/reality is actually inconsistent with your memories, because the reality/ dimension that was completely constant with your memories is the one you left because you died there. Before all this scientific tampering you probably seamlessly just merged without any noticeable differences that you could pinpoint. In that reality/dimension you would have died and could no longer remain there and so scientifically speaking you cannot stay to observe your own demise. According to my understanding of particle physics, no self-observing system can observe its own collapse without the observer affecting or altering the outcome.

So instead, you were to go to another constant yet comparable dimension/reality. However, now due to some (LHC and friends) interference you do not just go there seamlessly, but get somehow stuck between the dimensions even for just a few seconds at first. Here you still have your memories and just believe that you had a close call and everything is the same.

What if every close call in life, such as getting an small electrical shock plugging in a lamp, or almost getting hit by a semi-truck, or nearly falling from the roof of the building at work, was actually a death in an alternate reality/dimension and its not remembered because in the relocated reality/dimension it just never happened?

However being alive in the different reality/dimension doesn't mean death did not occur in the original one, so that in that reality/dimension all those close to you did experience the affecting sorrow and anguish of the death. There is a theory called Quantum Immortality. The notion being, if we die in one reality/dimension/time stream, we may just fuse into an alternate version of ourselves in another reality/dimension/time stream.

Spec·u·la·tion:
noun. ideas or guesses about something that is not known. the contemplation or consideration of some subject. a conclusion or opinion reached by such contemplation.

I do not **know** what any of this means; but, I will give you some of my thoughts that I think may have some probability or connection to the situation. The chief reason for this **speculation** is to attempt to understand the connection to events and alterations in our memories. Maybe we can get some idea of:

(1) What additional signs should we be looking for?
(2) What warnings and knowledge can we give for those who will be impacted on some level by even more these experiments?
(3) How should we investigate the correlation between timing of scientific events and coincidences with unexplained world events that may have created alterations of reality/dimensions?

Since these experiments there have been many changes that can only be explained as "abnormal" and "unprecedented." There will probably be a series of short-term, visible merging of present time with future time since the two coexist simultaneously. This may lead to periods of being able to predict things in the immediate future or to just know what will occur next as in a déjà vu.

Déjà. vu:
from French, literally "already seen", is the phenomenon of having the strong sensation that an event or experience currently being experienced has already been experienced in the past.

This could lead us to actually be stuck momentarily between two realities/dimensions observing other versions of ourselves in one or more realities/dimensions, (we would term it a precognitive dream) before finally being somehow propelled into the previously observed reality/dimension.

I do believe that reality and dimensions are still stable, however I am learning by experience that they can and do shift and change. I am now confident I am not living in my original timeline/ reality/dimension, nor with the original people I knew. For me it has much to do with a "physical" death and shift or change of consciousness being somehow interrupted or altered significantly in some way.

Do you think this is just imaginative thinking on my part? Well, think again. University of London physicist David Bohm, for instance, believes that objective reality may not be existent, that notwithstanding its seeming solid nature the universe is at its core an illusion, an enormous and superbly complete hologram. A hologram is a three-dimensional photograph made with the assistance of a laser. If a hologram of a flower is cut in half and then lit up by a laser, each half will still be found to contain the entire image of the flower. Undeniably, even if the halves are divided again, each piece of film will always be found to have a reduced but complete version of the original image. Unlike standard photographs, every fragment of a hologram comprises all the information possessed by the entire item and appears to be the item in complete form.

Bohm also considers the reason subatomic particles are able to remain in communication with one another irrespective of the expanse separating them is because their separateness may be an illusion.

He contends that at some profounder level of reality such particles are not individual entities, but *are really extensions of the same central something.* According to Bohm, the seeming faster-than-light communication between subatomic particles demonstrates a deeper level of reality and a more complex dimension beyond our own that is unknown to us.

So yes, if there are multiples versions of you, you could seem to be the complete version, while only being a part of a greater mass of the subatomic particles that make up yourself. If the apparent separateness of subatomic particles is deceptive, it means that at a deeper level of reality all things in the universe may be infinitely interconnected including each version of you in infinite dimensions living out various outcomes yet somehow all connected. It also means that there is communication faster than the speed of light!

It has been estimated that the human brain has the ability to remember somewhat on the order of 10 billion bits of information throughout the normal human lifetime. This is fascinating now because Stanford neurophysiologist Karl Pribram has also become convinced of the holographic nature of reality.

Pribram believes memories are programmed not in neurons, or minor assemblages of neurons, but in designs of nerve impulses. These impulses network the whole brain just like patterns of laser light interference intersect the complete piece of film holding a holographic picture. Pribram believes the brain is a hologram and his view has increasing support among fellow neurophysiologist. Even dreams and experiences involving "non-ordinary" reality can be understandable under the holographic example.

What this means for my theory is that in a universe in which individual brains are actually portions of a greater "hologram" and where everything is infinitely interconnected there could be multiple versions of ourselves with access to memories of the other versions of ourselves. Yes, this could scientifically help us understand the effects we are experiencing now termed the Mandela Effect.

Physicists recently announced that they expect to ultimately build larger and larger accelerators that would produce greater collisions with even more energy than the LHC. The proposed International Linear Collider would be more than 20 miles long rather than the ring design of the LHC at CERN. It's to be built in Japan and is to be operational by 2026.

Did you help to construct CERN's LHC? Were you even consulted? I'm pretty sure that you, just like me also had no part in it.

I definitely wasn't summoned to any of the summits and conferences in which the tactics and experiments were discussed and then agreed on; no matter what the consequences to the earth or its people.

Whether seen from a technical or a spiritual viewpoint, cautions have come from all sectors over resuming the CERN Large Hadron Collider experiments. First, you have scientists such as Stephen Hawking warning of possible universal devastation, while other scientists openly state to the public that they are seeking regardless of the consequences "other dimensions and parallel universes." If you think that sounds incredibly dangerous, you are not alone. Secondly, there are those of faith pointing to the misuse of Indian culture and spirituality with the placement of a statue of the god Shiva outside of the CERN facility in Switzerland. Vedic philosophy and culture have nothing to do with their stated aims of attempting to locate "dark matter," in an effort to destroy our world.

Numerous cautions from faith community members have been given and all are referencing the dangers of opening star gates, portals and wormholes and attempting to bring in entities that do not have access to this dimension. Yet they go unheeded or just mocked and ridiculed. No matter how much you have and how powerful may believe you are, there will be many things that will always be above and beyond the control of men and in the realm of the unseen.

Sadly we live in the age of ignorance and we call it intelligence and progress. These scientists who think that they are very smart want to play God, but lack the wisdom of God. Many notable scientists have said this technology could possibly destroy and it seems more than possible that a period of time could be either slowed down or sped up.

This machine may just be capable of distorting not only space, but also TIME. Another horrific thought to ponder is the possible universal and planetary consequences of emotionally damaged and ethically challenged international tyrants and sociopaths with unrestricted access to an instrument that can bend space and time. Men of power only want more power. History shows us sadly exactly how.

Chapter Four

Some Things Are Not The Same

When lines from movies, book titles, television show names and even people's names are not what we previously have known them to be, we could be seeing evidence of this multiverse right in front of us. When landmarks that never existed to you previously are now said to exist and are hundreds of years old, and when whole cities are completely different than you know it to have been it becomes very hard to just ignore it.

The way I see it, we seem to be able to shift in between different dimensions/realities and they don't seem like they are correct or "normal" somehow to our own experiences.

Although we have a very powerful and recognizably magnificent personal instrument; our Brain, which we still do not completely understand, we are willing to assume the "mistake" must be with us, and our thinking on some massive scale.

The world is a big, mysterious place, and it'd be foolish to believe we have all the answers, and now only these two choices. Our understanding is limited, but many of us who have experienced these things know something is radical altered and different.

Could the time streams of different parallel universe/ dimensions have been altered by the scientific experiments done previously? Could scientist somehow have fractured a time stream and now affect several dimensions by the circumstances occurring now with CERN?

As we all know there is NO WAY to prove the time stream has been altered, just as there seems to be no possibility of bringing or sending anything physical out of one dimension to another as evidence to prove a point.

Perhaps memories of that other time stream occasionally surfacing from time to time in our brains and multiple people experiencing this at the same time is the evidence of the change and the alteration.

For the past 5 or so years it seemed I was drifting between realities/parallel universes, without understanding what was happening. The only thing I knew for certain was that I didn't have that bad of a memory of my own existence. My recollections were precise and heavily detailed, and I was in severe shock on seeing things that I had already lived perceived differently. Yes, sometimes they were small things and it seemed unimportant, even minor but those little things still mattered to me and to those who remember differently.

There are many of us who are convinced that what we remember is real and we witnessed things that now we are told just never happened.

I definitely believe I was in a parallel universe and I detected a difference almost immediately, but I didn't understand what happened. The one great thing about this is that you can't use doubt to disprove the truth of someone else's memories.

It seems that the people from a different dimension/reality/time stream for some reason brought their memories intact with them. If you remember everything exactly as it is now you are in the dimension/reality in which you have always existed.

You might be surprised as to how effects like hyperspace, physical portals, and multi-dimensional existences can be affecting our daily lives. Yet this phenomenon has all become quite extensive since the CERN project in particular has been in operation. Once something is changed in the past, it changes the future. Entire things will be completely different; not partially different. When two dimensions collapse, or collide the dominant dimension seems to take over. Perhaps time streams were not just separated by time, but by particles of matter and these are the differences in physics itself, and somehow CERN has interfered with that.

Technically speaking, there would have been an infinite number of progressive dimensions, each individually divided by little alterations. But due to the fracturing of the time streams; events and things have changed and people are in strange and unfamiliar dimensions with strangers that look like the people they knew, and situations that are vastly different than their previous reality and recollections. What if CERN's LHC has inadvertently left remains of prior histories on current time streams, or has split entire time streams in half somehow?

Then there is ITER. This is the International Thermonuclear Experimental Reactor and is an enormous fusion reactor being constructed by 35 countries in southern France. ITER is also constructing a neutral beam test or NBFT site in Padova Italy. Strangely very few people will have heard of ITER or what they plan to do. While CERN actually began operations in 1954, ITER is still a decade away. In 2008 ITER and CERN signed a Cooperation Agreement to collaborate not only in the arenas of technology such as superconductors, electromagnets, cryogenics, control and data procurement and composite civil engineering, but also in organizational areas such as funding, purchasing and human resources. This collaboration includes software programs and working closely with DEISA, in full Distributed European Infrastructure for Supercomputing Applications; a European consortium of national supercomputing.

DEISA also maintains a network link with an agency known as Tera Grid – another supercomputing network in the US.

The ITER project has been branded as the world's paramount human attempt and illustration of world collaboration since the incident of the Biblical tower of Babel. This is so much more than it would at first appear; ITER has even created its own multi-national currency called the IUA. Why would a "science project" need its own currency?

The simple statement that a concept like the Large Hadron Collider with its stated aims even exists should be a warning signal, that evil is behind it. Human minds basically will not create a device such as this for no reason other than to "see if something happens."

If they are willing to spend billions of dollars on these huge contraptions, it must be something of an unimaginable magnitude that they are attempting to do.

I believe there are many parallel universes and perhaps they could in theory collide with each other and become somehow fractured, resulting in the transference of some type of magnetic energy and with it some of the people from one dimension/ reality into another without anyone realizing at first what might have occurred. There are many people who feel that CERN is modeled after the pineal gland or third eye because it's a vehicle for a variety of time travel mechanics.

There is of course another theory on why this is happening, and it is that it was intentional.

Maybe someone time traveled to aid in changing the future. They made slight changes because large changes would be too apparent and whoever controls or has authority over this sphere; be it some regime or clandestine social group would realize this had happened and could have possibly rectified it before any changes became observed. I have noticed that almost all the things that are changed are very public things that would go unobserved with the exception of ordinary regular people who have read, saw or utilized these things on a daily basis, and so would definitely have detailed recollections that would be hard to just ignore.

You cannot speak for anyone but yourself, so at the end of the day both recollections of the same event would be accurate, since it would be a personal experience. For me this is not philosophy, or speculation this is physics.

I know that the word "multiverse" is used instead of "universe" because scientists have accepted through quantum mechanics and the string theory that there truly are more than one dimension and more than one universe.

These are some things I have noticed in my original time stream of this multiverse that are completely different in this relocation one. See how many of them resonate with you as well.

Muhammad Ali is still alive (early 2015) but I recall his 2009 death, and the funeral on tv.

Frank Gifford died in 2012. His 2015 death was actually quite a surprise.

I Remember Betty White's death in 2014, It was announced that she died–"the last of the Golden Girls is Gone." Imagine my surprise that she is alive and well in 2016.

I remember Nelson Mandela having died in South Africa in prison and His wife Winnie Mandela later became the first black female president of South Africa.

I remember Scotland as being separate from England - its own island and was in the North Sea.

I remember having conversations with a family member that other family members say doesn't exist. Funny thing is he was at my wedding and brought a great wedding present.

Some people myself included remember Ronald Reagan dying in 1999, when he died again in 2004 in this 2015 time stream.

I learned in school that 2 bombs were dropped on Japan: Hiroshima, and Nagasaki however in other timelines it was 3 bombs.

I remember the death of Whitey Bulger in 2013, and a documentary on his life, yet in this time stream, he's still alive.

Mongolia is a gigantic country on global maps and a major world player in many people's reality. For me Mongolia was a province in China. Now in this reality time stream/dimension it is a large country between Russia and China.

I recall the company name as always "Proctor and Gamble" not "Procter and Gamble."

The comet that was the talk of our lifetime was called Hailey's Comet- Here it is Halley's Comet.

The Forrest Gump movie I saw when was first released is totally different than what we have here.

The hit song "Straight Up" by Paula Abdul sounds completely different today than I remember it when it was released.

There was actress Doris Day's dying in the late 2000's yet she is still alive now.

Gone with the Wind's Scarlett O'Hara's famous line: "Wherever Shall I go; whatever shall I do," is here stated differently.

Both my daughter and I remember Cheesecake Factory restaurant meals jokes about the food being horrible and the joke was you just ordered the food to stave off the sugar shock from the magnificent desserts.However, in this time stream it is considered a great place for a good meal and people rave about the food.

I remember being taught about the 52 states. Some children born after 1990 seem to all remember 50 states.

I recall John Goodman's death from a heart attack shortly after the Flintstones movie in 1994, but in this time stream he has lost weight and is alive in 2015.

 I remember Korea being SOUTH of China near Vietnam, certainly not out North next to Eastern Russia.

I remember that the Lindbergh baby had never been found, yet here it is reported as having been found dead.

The movie line in the Wizard Of Oz was "Toto, I don't think we're in Kansas anymore." In this time stream it is "Toto, I have a feeling we're not in Kansas anymore."

Reba McIntyre is spelled McENTIRE now- which is different from the Scottish ancestry of McIntyre in my remembered time stream.

I remember Vladivostok being much more North in Russia and not bordering Korea as it is shown on maps in this time stream.

I remember watching on television that event in Tiananmen Square in which the Chinese young man refused to get out of the way of the army tank and was run over and killed. It was shocking and everyone was speaking about it. In this time stream that never happened.

 Some remember Fruit Loops breakfast cereal in the 1960s, yet I only remember it first appearing in stores in the late 1970s. And it was always spelled Fruit Loops, not FROOT LOOPS as in this time stream.

I remember the air and fabric freshener product Febreeze, in this dimension it is Febreze.

In my original time stream/ dimension there was no animal called a Narwhal. This is apparently a whale with a long horn on its head like the unicorn. In this time stream it is called "the unicorn of the sea." Growing up I watched nature programs on TV–Jacque Cousteau , Marlin Perkins, David Attenborough, and others and never heard of this. I thought it was some kind of a joke since I've never heard of nor seen a picture of a narwhal before 2016.

I recall Easter Island as having been discovered by James Cook/ Easter Island, and I remember him finding it uninhabited. Rapa Nui is the name of what I knew as Easter Island, given to it by its native people, who have continually inhabited the island for nearly 3,000 years in this time stream/dimension.

I remember the sun being bright yellow, not white, and I learned in science classes at school that there were only 4 or 5 cloud formation types , but in this time stream clouds appear in odd shapes and forms and there are over 20 types here.

Thanksgiving was always on the third Thursday of November in the United States. And in this time stream, it's the fourth Thursday in November. It stands out in my mind because my grandmother taught it to me as a child, when I learned the countdown to Christmas day.

 I remember the peace sign become popular in the 1970s; it had the arms facing upward; never downwards as in this time stream.

I remember the GREAT Pyramid of Giza being off into the desert MILES away not literally 700 FEET from the suburbs of the city of Cairo Egypt, as it is here.

 I remember that Jane Goodall died and was remembered for her research on gorillas, when in this time stream she is still alive and famous for her research with chimpanzees.

Gorillas in the Mist was a movie which had a TV premier and I distinctly remember the movie I saw was about Jane Goodall; staring Susan Sarandon. In this time stream Sigourney Weaver is the actress in that movie and it is about Diann Fossey.

I remember the pictures of this massive white statue called Christ, the Redeemer overlooking the city of Rio de Janeiro on a gigantic white rectangular base. Now it is just a large statue. The base has also radically and mysteriously changed to a smaller base and is a black square cube.

I remember a BBC America Television show called MI-5, however in this time stream it is called SPOOKS, and while still about MI-5, it was never called that especially in the US.

Cartoons were Looney Toons now Looney Tunes and Merrie Melodies in this time stream, yet I knew it as Merry Melodies my entire childhood.

I remember a peanut butter known as "Jiffy" the original brand name. So when I saw "Jiff" peanut butter I thought it was a name change by the company. It seems that at least in this time stream there was never a name change and it has always been known as "Jif." However I remember my brother and I being very insistent with my mom when we were children that she only buy "Jiffy" and not "Skippy" another brand of peanut butter. I even remember the song from the commercial.

I remember Oscar Meyer as a deli product company, in this time stream it is Oscar MAYER. I even remember singing the song in the commercial...about "my bologna has a first name it's ----O-S-C-A-R, my bologna has a second name it's----- M-E-Y-E-R....!"

I also recall that all traffic lights were green yellow and then red at the bottom, so I was surprised when I noticed it in reverse.

I also remember the spelling of words being completely different. I spelled a word as "suprise" now it's "surprise" and "lightening" instead of "lightning", and "realise" is now "realize." I was always big on reading and writing and had entered spelling contests every year as a child. I paid attention to words and I am a writer now, so I find this bizarre. We were taught the proper grammatical usage is "my brother and I," now here in this time stream it is "my brother and me."

Here combined words are non-existent: 'infact' is now "in fact"; afterall to "after all"; overall to "over all" moreso to "more so;" alot to "a lot"; alright to all right; and no-one is "no one."

"Dilemma" is remembered as being spelled "dilemna" and "dammit", as "damnit" The spelling of the nation of Columbia changed to Colombia.

The colors chartreuse and puce have switched here. I remember Chartreuse a pink -reddish purple, not puce's yellowish-green color.

The automobile symbols are different also. Volkswagen – VW, here has a space between the two monogram letters, and Volvo in this time stream has an arrow added to the circle, making it the symbol for "male,"and not the circle missing a piece that I remember.

Vancouver Island seems larger here and British Columbia is much larger also on these maps.

I remember New Zealand being one land mass. In this time stream it is now two islands and it is bigger than Italy.

The Bahamas were NEVER just off the coast of Florida in my time stream, only Bermuda was. Cuba was NEVER that close to Mexico. Also there was no island off the coast of Cuba!

When I visited NYC years ago Manhattan Island jutted out into the Atlantic. The statue of Liberty was on an island a little farther out into the Atlantic and not near New Jersey. You had to take a ferry to get to Staten Island as they never had a bridge.

Here in this time stream I learned there are 4 bridges to Staten Island.   I had no idea that there was any bridge. I always thought that you had to use a ferry to go to Staten Island.

I do remember in the movie Working Girl, actress Melanie Griffith had to ride the ferry back to Staten Island and I clearly recall the scene. In this time stream the movie does not have that scene.

Martha's Vineyard was a district on Long Island. It has been moved away, leaving the Bay in Long Island, here Martha's Vineyard is an island.

I recall Sri Lanka being directly South of India, not off to the East of it. I was shocked to see Gibraltar moved from the strait between Spain and Morocco –to be on the East coast of Spain.

I was stunned in particularly by South America's 1000 mile eastern shift, out of what I recall as the straight alignment with North America.

The JC Penny Store in this time stream is JC Penney.

American Television chef of "Bizarre Foods" was Andrew Zimmerman, here he is Andrew Zimmern.

The host of the Twilight Zone television series was known as Rod Sterling, here he is Rod Serling.

Walmart was ALWAYS a blue logo- never Wal*mart in white logo in my original dimension.

The Talladega Superspeedway and the Daytona Raceway were in Florida, however in this dimension the Talladega Superspeedway is in ALABAMA. I was shocked that there isn't even a town called Talladega in Florida.

Then again there is that Rock of Gibraltar. It is British owned. It is in my time stream/Dimension, a source of contention between England and Spain because Spain believes due to its proximity to their coast it should be considered as Spanish territory even though it is an island offshore. This is how I remember it; an island of disputed territory, not a part of the land mass of the country of Spain, sitting surrounded by water facing Morocco.

I have the same memories and many more as do some of my friends and family; it seems that almost everything is incorrect, countries and landscapes, words, people in the public eye and often historical events from what I have always known. If you don't recall things in this way, you are from THIS dimension/reality/ time stream, or, alternately from yet ANOTHER alternate dimension/reality/time stream from my original one.

My belief is that these are dimensional shifts or fractures that we were experiencing are variations in reality which is different than our memories. Some horrible technological tampering has done something peculiar and it is affecting everyone, some I believe just haven't noticed yet, or are putting it down to being mistaken and shrugging it off.

It may be that two or more dimensions have collided and somehow like glass tubes they are fractured no longer sealing in one reality from another right next to it.

It's scientifically INEVITABLE that when a large group of people remember something they knew for certain, it's true for them. And I mean by a large group of people – MILLIONS OF PEOPLE, who have experienced the scenario. I believe that there must be something more here than meets the eye.

When people say the difference in what I remembered is just a memory lapse I realize their dismissive attitude is meaningless when MILLIONS of other people also recall it differently and with the same miniscule details that I do.

I am still amazed over the geographical changes on the world map. It bothers me immensely as I vividly remember a land mass being called the North Pole it was never a large lake. New Zealand was above Australia and it was one land mass and not in two pieces.

Australia is now half the size I remember and is missing part of its shape at the top. Indonesia and Australia are much closer to each other in this time stream. Australia in my original dimension/time stream reality was out in the ocean completely isolated and very far from any other land mass.

It happened to me again, my reality shifted I believe around 2013, in the midst of continuous disturbing personal activities.

However even before that time, I remember in 2008 waking up one morning and I got out of bed and went to my daughter's room and stared at her while she slept to make sure "she was still there" and I went back to bed. I woke up some time later, seemingly okay and I wondered "now why did I do that?" I did not understand but I felt that something was very wrong and I was uneasy for many days after that.

I do remember the sun use to be brighter before that day and the sky used to be very blue and crystal clear not opaque and dull. Perhaps I was moved somehow, to stop the events that were happening; because of the stress ordeal and other related issues in my life. But the how and why I have no idea. I am still asking myself why it is so different this time.

My own personal time stream has changed drastically since 2001. We were always close and seemingly overnight there were deep issues with my immediate family, and complete isolation from one another.

I remember being very scared all the time then because I felt that "this was not my life." When my reality somehow changed my memories seemed different to what I was seeing after 2001. Everything seemed different and even friends seemed not the same people I knew before. I continued to say out loud to no one in particular that "sometimes I just don't resonate with these humans." When I'm feeling as though I'm not like the people around me, I can also feel the strong difference when I met a "regular" person like myself. There may be some validity to something having happened when you hear of people who seemed "out of character" and then they disappeared out of daily life without a trace.

I also have distinct memories of three near death incidences where I either just woke up, found myself outside on the ground, or realized I was sitting somewhere and did not know what happened as I did not recall how I got there but I thought "at least I'm ok now." Maybe I've died at least three or more times already, and with each death I wake up in a different dimension/reality.

The idea that Einstein is actually right; and that when you match the frequency of the reality/dimension you want you go into, then that reality/dimension becomes your new reality/dimension is really quite astounding when it may be happening to you.

Now I know for absolute certain that something is seriously going on. I feel the necessity to "detail check" with my daughter and close friends to see if they continue to share my reality dimension/time stream memories. At first I tried to ask subtle questions without drawing too much attention to why I was inquiring about the simplest things. Now for some time I have also had the thought that my reality may be unlike someone else's reality.

Many people have had strange, paranormal experiences with a piece of clothing missing, or being a different size or color or just popping up in strange places. You simply cannot just write all these realities off as the result of massive bad memory, or brain lapses due to age or stress.

So many people can recall funerals of famous people only to learn they are alive and didn't die. I feel it's because some of us have moved into a different universe/dimension and time stream where these people haven't died yet.

That then explains why some of us remember deaths while others do not. We have no explanations, but there are many suspicions.

The cause of all of this could be the hadron collider at CERN and its work on dimensional portals. The scientists have ignored the fact that when something changes, it changes not just here or there, but EVERYWHERE.

You enter a completely new time stream of consciousness where the whole past is different. Consequently when you go back into the past, it's has completely changed.

That's why there are no newspaper clippings, paintings, advertising commercials, and even your own DVD movie collection that you can access to verify your own memories. It has all changed when you did and the possibility exists that you may have done more than physically move.

 Any two realities that share a timeline close enough can stream-slip together. Anything that is the same between the two will merge. There is so far no way to say how that is determined. It's apparent though, because if someone was reminisced as dead but realized to be alive, undoubtedly the reality dimension/time stream on which the individual is still alive has taken dominance. Thinking through even my own rationalizations for what I have experienced, I feel sometimes as though someone has tried to trick or program me, or as though I'm in some type of sick mental experiment. To me reality is so unreal and as I get older this feeling intensifies. So for anyone who's new to this idea — and slightly astounded, identifying many things as different now – yes, it's okay to feel unstable. For me the question isn't whether this happens, but how it's happening and why.

Other Books By Roshan Cipriani

Rise - Be True To Yourself-Inspire Others To Live
How To Get Through Any Wall In Your Life
Train Up A Child – A Scriptural Guide To Parenting
The Art Of War For Parenting Your Teenage Child- How To Win A War You Didn't Even Know You Were In
The Key To This Life - Conscious Faith In An Unconscious World
Destiny – Past Present Future
Life Lessons Learned
In The Fire – Accessing Miracle Power During A Crisis
The Kingdom Lifestyle - Living By Faith And Not By Sight
God's Secret Wisdom –Principles And Secrets Of The Kingdom Of God
The Greatest Principle - The Kingdom Of God And Biblical Economics
Bricks Without Straw- Spoiling Egypt And Spoiling Babylon; The Mighty Wealth Transfer
When Failure is Not An Option
Real Faith – How To Have It And Why It Matters
The Bibles Healing Promises
I Say What They Said- Miracle Bible Prayers
The Psychology Of Stress-Dismantling The Enemy's Weapon Now
Never Quit-The Secret To Getting Through Any Wall In Your Life
The Seventy Two Lunar Sabbaths- Sabbath Observance By The Phases Of The Moon
BUSINESS PLAN: Make God Your Partner –He Commanded His Blessings
PROSPERITY CONSCIOUSNESS – Living In An Abundant Universe (Personal Biblical Economics) Volume 1
Metamorphosis-Mirrors Of The Soul, Awakening To The Real You
Waiting In Goshen
How To Be Smart And Have Common Sense
None Of These Diseases –Sickness And Genocide In Second Egypt
DIVORCE RECOVERY: How To Live Again
UFO COVER-UP: Biblical Evidences Uncovered-(Conspiracy) Volume 1
TRAVEL: How To Behave On An Airplane
Living In A Fractured Multiverse-The Reality Shift Effect
Second Exodus From Second Egypt –Volume 1
Asset Protection And Wealth Management-Volume 1 -Trust And LLC For Legal Asset Protection
PETS LIFE AFTER LIFE-Assurance Of Your Reunion
RELATIONSHIP RESCUE FROM THE BIBLE: What The Bible Says About Relationships
LETTING GO OF PHARAOH-Preparations For Future Crisis
Second Exodus From Second Egypt – Volume 2
BONES OF CONTENTION: The Coming Resurrection Of The Super Nation Of Real Israel
AWAKE IN THE DARK:*ON THE ROAD FROM BABYLON TO ZION*
THE LOST SHEEP OF THE HOUSE OF ISRAEL
THE LOST HISTORY OF THE WORLD – Volume 1
ASSET PROTECTION 2: Wealth Management For Global Living
DISCERNMENT: The Awakening Of Real Israel
TO KNOW OURSELVES ONCE AGAIN: Your Future In Real Israel
Last Days Wisdom: FOR REAL ISRAEL
Scary Close Prophecy
AFTER BABYLON
REAL Prophecy For REAL Israel

www.ingramcontent.com/pod-product-compliance
Lightning Source LLC
Chambersburg PA
CBHW080716190526
45169CB00006B/2395